.ecture Notes in Mathematics

informal series of special lectures, seminars and reports on mathematical topics

ited by A. Dold, Heidelberg and B. Eckmann, Zürich

Petru L. Ivănescu

Institute of Mathematics, Romanian Academy of Sciences, Bucharest

Pseudo-Boolean Programming and Applications

Presented at the Colloquium on Mathematics and Cybernetics in the Economy, Berlin, October 1964

965

Springer-Verlag · Berlin · Heidelberg · New York

Abstract of the author's Thesis, for which he obtained a Doctor's Degree from the University of Bucharest in June 1965.

Library of Congress Catalog Card Number 65-29241. Printed in Germany. Title No. 7329

Introduction

In this lecture we shall give a brief description of some applications of Boolean algebraic methods in operations research.

It was Professor George B. Dantzig, who has shown (/22/, /23/, Chapter 26 of /24/, etc.) that a great variety of problems in operations research and related areas, may be given a unified approach by means of mathematical programming with bivalent variables.

The initial idea concerning the possibility to apply Boolean methods in economic problems, belongs to Professor Robert Fortet, who, in /35/ has pointed out the strong relationship between Boolean algebra and some important combinatorial problems arising in operations research. In /36/ he has also shown that problems of real economic importance may be handled with that mathematical apparatus, while in /37/ he used it in connection with problems of linear programming in integers. Important results in this direction were elaborated by Paul Camion /19/ and Khaled Maghout /89/.

The aim of this lecture, however, consists only in presenting the results obtained by the author in collaboration with Sergiu Rudeanu, Ivo Rosenberg and Aristide Deleanu.

Our attention was drawn on the topic in discussion by two major facts.

The first was our close affiliation with the research team lead by Professor Gr. C. Moisil, whose important contributions on the applications of Boolean algebras and Galois fields in switching theory (see for instance /93/,/94/,/95/,/96/,/97/,etc.)

were of an outstanding importance. It was natural to look for the economic utility of a tool which proved itself so useful in techniques.

Secondly, studying different questions related to the transportation problems of operations research (/4/,/5/,/6/, /7/,/8/,/9/,/1o/,/11/[3],/12/[3],/13/,etc.) we noticed that in the method of Egervary /29/ for solving the problem, one important step was not given a systematic solution and that, this step may be solved easily /73/,/74/,/75/) with Boolean techniques.[4] In fact, it was in this context, that the method of pseudo-Boolean programming has been employed, although at the moment it was not given a general description.

This paper will be made up of two parts. In the first the method of pseudo-Boolean programming will be described together with its use in solving problems of integer polynomial programming. The second part will contain a presentation of some applications of this method to combinatorial-type problems arising in operations research and related fields. The paper will be concluded by a list of problem we have now in research.

In order to keep the paper within a moderate length, we will give here no proofs; for proofs and details the reader is refered to the appropriate references indicated in the text. We have tried to give a fairly complete bibliography.

---

3) In connection with /11/ and /12/ see also paper /21/ by A. Charnes and M. Kirby.

4) Another solution of this problem was given by B. Krekó /84/, and the question was also studied by I. Kolumban /78/.

P A R T I

Chapter I

Notations and Terminology

§ 1 Boolean Algebras with two Elements

Let us introduce in the two-element set made-up only of 0 and of 1, besides the usual arithmetical operations of addition ("+"), substraction ("-") and multiplication ("." or simply juxtaposition), the bynary operation of union ("$\cup$"), defined by

$\cup$	0	1
0	0	1
1	1	1

and the operation of negation (" $\bar{}$ ") defined by

a	0	1
$\bar{a}$	1	0

It is easy to notice that

$$a \cup b = a + b - ab \qquad (1.1)$$
$$\bar{a} = 1 - a$$

and that the operation of union is commutative and associative. This properties entitle us to introduce the symbol

$$\bigcup_{\ell=1}^{k} a_\ell = a_1 \cup a_2 \cup \ldots \cup a_k; \qquad (1.2)$$

of course, we shall always consider

$$\bigcup_{i \in \emptyset} a_i = 0 \qquad (1.4)$$

and

$$\prod_{i \in \emptyset} a_i = 1; \qquad (1.4')$$

here $\emptyset$ stands for the empty set.

For simplifying notation, we sometimes put $x \in \{0,1\}$, $a \in \{0,1\}$:

$$x^a = \begin{cases} x & \text{if } a = 1 \\ \bar{x} & \text{if } a = 0 \end{cases} \qquad (1.5)$$

The set $\{0,1\}$ with the operations of union, multiplication and negation is a Boolean algebra[5)] and will be denoted with L_2.

A finite expression made up only of the constants of L_2, and variables of L_2 with the aid of the operations of union, multiplication and negation will be called a <u>Boolean expression</u> (BEX).

Using formulas (1.1), (1.2) and its generalizations

$$a \cup b \cup c =$$
$$= (a \cup b) \cup c = (a+b-ab) + c - (a+b - ab)c =$$
$$= a + b + c - ab - bc - ca + abc, \qquad (1.6)$$
$$a \cup b \cup c \cup d =$$
$$= a+b+c+d - ab - bc - cd - da + abc + bcd + cda + dab-abcd, \qquad (1.7)$$

etc., we may transform any Boolean expression into one, written only with the aid of the arithmetical operations of addition, substraction and multiplication; the expression resulted in this way from the Boolean expression $\mathcal{E}$ will be called its <u>arithmetical equivalent</u> and denoted with $\mathcal{E}^+$.

If $\mathcal{E}$ and $\mathcal{F}$ are two BEX, any one of the following relations

$$\mathcal{E} = \mathcal{F} \quad \text{(Boolean equations = BET)}$$

$$\left.\begin{array}{l} \mathcal{E} \leq \mathcal{F} \\ \mathcal{E} \geq \mathcal{F} \\ \mathcal{E} < \mathcal{F} \\ \mathcal{E} > \mathcal{F} \end{array}\right\} \text{(Boolean inequalities = BIN)}$$

5) For the definition and properties of Boolean algebras see for instance G. Birkhoff /18/, M. Denis-Papin, A. Kaufmann, R. Faure /27/, H.G. Flegg /31/, G. Szasz /113/, etc.

will be termed a Boolean relation (BR).

§ 2 Pseudo-Boolean Functions

The Chartesian product $L_2 \times L_2 \times \ldots \times L_2$ will be denoted by L_2^n, i.e.

$$L_2^n = \left\{ (c_1, \ldots, c_n) \;\middle|\; c_i \in L_2 \qquad (i=1,\ldots,n) \right\} \tag{1.8}$$

Any application F of L_2^n in the ring Z of the integers

$$F : L_2^n \longrightarrow Z \tag{1.9}$$

will be called a pseudo-Boolean function (PBF)[6]. That means that the arguments of a PBF may take only the values 0 or 1, while its values are integers.

If in the definition of Boolean relations we replace the BEX by PBF, we obtain pseudo-Boolean relations (PBR).

It is not difficult to prove (see /7o/)

THEOREM 1. Any PBF may be written as a polynomial with integer coefficients, linear in any of its variables.

Pseudo-Boolean functions and relations occur in many a problem of operations research. Numerous examples can be found in chapter 26 of /24/, in /36/, as well as in the subsequent chapters of this paper. As an illustration we quote a simplified version of the well-known cargo-loading problem (see, for instance /16/): a vessel is to be loaded with a cargo composed of different type if items: 1,2,...,n. Denoting with v_i the value of the i th type of item, with w_i its weight, and with z the maximum capacity of the vessel, the problem consists in determining the most profitable cargo of the ship, i.e. in determining

$$(x_1, \ldots, x_n) \in L_2^n$$

---

6) R. Fortet terms these functions: "fonctions algébriques entières".

so that respecting the PBIN

$$w_1x_1 + \dots + w_nx_n \leq z, \tag{1.10}$$

the PBF

$$v_1x_1 + \dots + v_nx_n \tag{1.11}$$

would be maximized. (We have supposed without loss of generality that v_i, w_i and z are integers).

§ 3 Galois Fields Modulo 2

Professor Gr. C. Moisil has shown (see for instance /96/, /97/) that the use of Galois Fields is sometimes more advantageous in switching algebra, then that of Boolean algebras. P. Camion utilized in /19/ this apparatus for problems of operations research. In /55/, /56/, /63/, we have applied it for solving problems of integer polynomial programming.

Let us introduce in the set $\{0,1\}$ the operations of "sum modulo 2" ("Δ") defined by

Δ	0	1
0	0	1
1	1	0

It is easy to notice that,

$$a \Delta b = a + b - 2ab \tag{1.12}$$

$$= a\bar{b} \cup \bar{a}b \tag{1.13}$$

and that

$$\bar{a} = a \Delta 1 \tag{1.14}$$

$$a \cup b = a \Delta b \Delta ab \tag{1.15}$$

$$= a \Delta \bar{a}b \tag{1.16}$$

$$= a\bar{b} \Delta b$$

The operations of sum modulo 2 being commutative and associative, we may introduce the symbol,

$$\underset{i=1}{\overset{k}{\Delta}} a_i = a_1 \Delta \ldots \Delta a_k \qquad (1.17)$$

and we consider

$$\underset{i \in \emptyset}{\Delta} a_i = 0 \qquad (1.18)$$

The set $\{0,1\}$ with the operations of sum modulo 2 and multiplication is a field, and it will be called the "Galois Field Modulo 2" and denoted GF(2).

Galoisian expression (GEX) and relations (GR) are definied analogously with BEX and BR. Equality between Galoisian expressions is usually denoted with "$\equiv$" and termed "congruence modulo 2".

Chapter 2

Minimization of Pseudo-Boolean Functions [7)]

§ 1 Description of the Algorithm

Let us consider the problem of determining those

$$(x_1, \ldots, x_n) \in L_2^n \qquad (2.1)$$

for which the PBF

$$F_1(x_1, \ldots, x_n) \qquad (2.2.1)$$

attains its minimum.

Using theorem, 1, we see that we may write

$$F_1(x_1, \ldots, x_n) = x_1 g_1(x_2, \ldots, x_n) + h_1(x_2, \ldots, x_n) \qquad (2.3)$$

where g_1 and h_1 are PBF of $x_2, x_3, \ldots, x_n$.

Let us denote

$$M_1 = \left\{ (a_2, \ldots, a_n) \in L_2^{n-1} \,\middle|\, g_1(a_2, \ldots, a_n) < 0 \right\} \qquad (2.4.1)$$

7) For proofs and details see /70/, /72/.

$$N_1 = \{(b_2,\ldots,b_n) \in L_2^{n-1} \mid g_1(b_2,\ldots,b_n) = 0\} \tag{2.5.1}$$

and put

$$x_1 = \bigsqcup_{(a_2,\ldots,a_2)\in M_1} x_n^{a_2}\ldots x_n^{a_n} \cup u_1 \left[\bigsqcup_{(b_2,\ldots,b_n)\in N_1} x_2^{b_2}\ldots x_n^{b_n} \right] \tag{2.6.1}$$

where u_1 is an arbitrary parameter in L_2.

Let us denote with x_1^o the expression of x_1 obtained by taking $u_1 = 0$ in (2.6.1):

$$x_1^o = \bigsqcup_{(a_2,\ldots,a_n)\in M_1} x_2^{a_2}\ldots x_n^{a_n} \tag{2.6.1^o}$$

and with x_1^+ the arithmetical equivalent of x_1^0 :

$$x_1^+ = \sum_{(a_2,\ldots,a_n)\in M_1} x_2^{a_2}\ldots x_n^{a_n} \tag{2.6.1$^+$}$$

where this time every $\bar{x}$ will be replaced by $1-x$.

Let us now put

$$F_2(x_2,\ldots,x_n) = F_1\left[x_1^+(x_2,\ldots,x_n),\ x_2,\ldots,x_n\right], \tag{2.2.2}$$

$$F_2(x_2,\ldots,x_n) = x_2 g_2\ (x_3,\ldots,x_n) + h_2(x_3,\ldots,x_n) \tag{2.3.2}$$

(with g_2 and h_2 being PBF of $x_3, x_4,\ldots,x_n$),

$$M_2 = \{(a_3,\ldots,a_n) \in L_2^{n-2} \mid g_2(a_3,\ldots,a_n) < 0\} \tag{2.4.2}$$

$$N_2 = \{(b_3,\ldots,b_n) \in L_2^{n-2} \mid g_2(b_3,\ldots,b_n) = 0\} \tag{2.5.2}$$

and

$$x_2 = \bigsqcup_{(a_3,\ldots,a_n)\in M_2} x_3^{a_3}\ldots x_n^{a_n} \cup u_2 \left[\bigsqcup_{(b_3,\ldots,b_n)\in N_2} x_3^{b_3}\ldots x_n^{b_n} \right] \tag{2.6.2}$$

(where u_2 is an arbitrary parameter in L_2),

$$x_2^+ = \sum_{(a_3,\ldots,a_n)\in M_2} x_3^{a_3}\ldots x_n^{a_n} \qquad (2.6.2^+)$$

etc.

Generally,

$$F_i(x_i,\ldots,x_n) = F_{i-1}\left[x_{i-1}^+(x_i,\ldots,x_n),\ x_i,\ldots,x_n\right] \qquad (2.2.i)$$

$$F_i(x_i,\ldots,x_n)=x_i g_i(x_{i+1},\ldots,x_n)+ h_i(x_{i+1},\ldots,x_n) \qquad (2.3.i)$$

(with g_i and h_i being PBF of $x_{i+1},\ldots,x_n$),

$$M_i = \left\{(a_{i+1},\ldots,a_n)\in L_2^{n-i} \,\middle|\, g_i(a_{i+1},\ldots,a_n) < 0\right\} \qquad (2.4.i)$$

$$N_i = \left\{(b_{i+1},\ldots,b_n)\in L_2^{n-i} \,\middle|\, g_i(b_{i+1},\ldots,b_n) = 0\right\} \qquad (2.5.i)$$

and

$$x_i = \bigcup_{(a_{i+1},\ldots,a_n)\in M_i} x_{i+1}^{a_{i+1}}\ldots x_n^{a_n} \cup u_i \left[\bigcup_{(b_{i+1},\ldots,b_n)\in N_i} x_{i+1}^{b_{i+1}}\ldots x_n^{b_n}\right]$$

(2.6.i)

(where u_i is an arbitrary parameter in L_2),

$$x_i^+ = \sum_{(a_{i+1},\ldots,a_n)\in M_i} x_{i+1}^{a_{i+1}}\ldots x_n^{a_n} \qquad (2.6.i^+)$$

Finally, we obtain

$$F_n(x_n) = x_n g_n + h_n \qquad (2.3.n)$$

where g_n and h_n are constants (in Z).

We put now

$$x_n = \begin{cases} 1 & \text{if the constant } g_n < 0 \qquad (2.6.n.1)\\ 0 & \text{if the constant } g_n > 0 \qquad (2.6.n.2)\\ u_n & \text{if the constant } g_n = 0 \qquad (2.6.n.3)\end{cases}$$

u_n being an arbitrary paramter in L_2.

From (2.6.n) we see that x_n is a function of u_n:

$$x_n = X_n(u_n) \tag{2.7.n}$$

(which, of course, may be a constant).

Introducing (2.7.n) in (2.6.n-1) we get x_{n-1} as function of u_n and u_{n-1}:

$$x_{n-1} = X_{n-1}(u_{n-1}, u_n). \tag{2.7.n-1}$$

Analogously, we obtain

$$x_{n-2} = X_{n-2}(u_{n-2}, u_{n-1}, u_n) \tag{2.7.n-2}$$

....

$$x_i = X_i(u_i, u_{i+1}, \dots, u_n) \tag{2.7.i}$$

....

$$x_1 = X_1(u_1, u_2, \dots, u_n) \tag{2.7.1}$$

The following theorem is proved:

THEOREM 2 For any particular system of values, of the parameters $(u_1,\dots,u_n) \in L_2^n$, the system (2.7.n), ..., (2.7.1) yields a system of values $(x_1,\dots,x_n) \in L_2^n$ for which F_1 attains its minimum; conversely, for any system $(x_1,\dots,x_n) \in L_2^n$ minimizing F_1, there is a system $(u_1,\dots,u_n) \in L_2^n$ so that the relations (2.7.n),...,(2.7.1) take place.

Introducing the values

$$x_n' = \begin{cases} 1 & \text{if} \quad g_n < 0 \\ 0 & \text{if} \quad g_n \geqslant 0 \end{cases} \tag{2.6.n'}$$

in (2.6.n-1⁺) we obtain a value x_{n-1}'; introducing x_n' and x_{n-1}' in (2.6.n-2⁺) we obtain a value x_{n-2}', etc. It is proved that

THEOREM 3 The system of values $x_n', x_{n-1}', \dots, x_1'$ minimizes F_1.

In case we are interested only in one minimum of F_1 we may use theorem 3 which do not require the computation of $N_1, N_2, \dots, N_n$. However, if we wish to obtain all the values minimizing F_1 we have to apply theorem 2, and for this the computation of $M_1,\ldots,M_n$, $N_1,\ldots,N_n$ is required.

<u>Note 1</u> If $g_h(x_{h+1},\ldots,x_n)$ does actually depend not on all of its arguments, but only on $(x_{i_{h+1}},\ldots,x_{i_n})$, then we may take in (2.4.h), (2.5.h) and (2.6.h), M_h, N_h^n and x_h depending also only on $(a_{i_{h+1}},\ldots,a_{i_n})$, and the corresponding $(x_{i_{h+1}},\ldots x_{i_n})$

<u>Note 2</u> The computation of $x_1,x_2,\ldots,x_n$ may be carried out in an order different from the above, if this seems to be more convenient. The following approach may be suggested: Let $o_j(x_i)$ be the number of terms of F_j containing x_i. Begin with computing that x_k for which $o_j(x_k)$ is smallest.

§ 2 <u>An Example</u>

Let us determine the minima of

$$F_1 = 2\,x_1 + 3\,x_2 - 7\,x_3 - 5\,x_1x_2x_3 + 3\,x_2x_4 + 9\,x_4x_5$$

Here

$$g_1 = 2 - 5\,x_2x_3,$$

and

$$M_1 = \{(1,1)\},$$
$$N_1 = \emptyset$$

and hence

$$x_1 = x_2^1\,x_3^1 = x_2x_3 \qquad (2.8.1)$$

$$F_2 = -3\,x_2x_3 + 3\,x_2 - 7\,x_3 + 3\,x_2x_4 + 9\,x_4x_5,$$

$$g_2 = -3\,x_3 + 3 + 3\,x_4,$$

and

$$M_1 = \emptyset$$
$$N_1 = \{(1,0)\},$$

and hence

$$x_2 = u \,.\, x_3^1 \,.\, x_4^0 = u \,.\, x_3 \,.\, \bar{x}_4 \,. \qquad (2.8.2)$$

$$x_2^+ = 0. \qquad (2.8.2^+)$$

$$F_3 = -7\, x_3 + 9\, x_4 x_5,$$

$$g_3 = -7,$$

$$x_3 = 1 \qquad (2.8.3)$$

$$F_4 = -7 + 9\, x_4 x_5,$$

$$g_4 = 9\, x_5$$

$$M_4 = \emptyset$$

$$N_4 = \{(0)\},$$

and hence

$$x_4 = v x_5^0 = v \bar{x}_5 \qquad (2.8.4)$$

$$x_4^+ = 0 \qquad (2.8.4^+)$$

$$F_5 = -7,$$

$$g_5 = 0,$$

$$x_5 = w. \qquad (2.8.5)$$

From (2.8.5), (2.8.4), (2.8.3), (2.8.2), (2.8.1) we have:

$$\left.\begin{aligned} x_5 &= w. \\ x_4 &= v\, \bar{w} \\ x_3 &= 1 \\ x_2 &= u(1-v + vw) \\ x_1 &= u(1-v + vw) \end{aligned}\right\} \qquad (2.9)$$

where u, v, w are arbitrary parameters in L_2

For the different values of $(u,v,w) \in L_2^3$ we find all the minima of F_1 :

$$\left.\begin{aligned} F_1(0,0,1,0,0) &= \\ F_1(0,0,1,0,1) &= \\ F_1(1,1,1,0,0) &= \\ F_1(1,1,1,0,1) &= \\ F_1(0,0,1,1,0) &= \end{aligned}\right\} = -7$$

Chapter 3

Systems of Pseudo-Boolean Equations and Inequalities

In this chapter we shall give three alternative methods for solving systems of PBET and PBIN. They were described in /55/, /56/, /59/, /60/61/64/. They extend the well-known results of W.E.Johnson /77/, L.Löwenheim /86/, S.Rudeanu /108/,/109/,/110/ for the solution of Boolean equations to that of pseudo-Boolean ones.

§ 1 A Direct Approach

Let us consider a PBIN, for instance

$$f(x_1,\dots,x_n) \leqslant 0 \tag{3.1}$$

and let A be the sum of the negative coefficients of a polynomial expression of f; denoting with a the smallest integer with the property that $-A \leqq 2^a$, let us form the PBET

$$f(x_1,\dots,x_n) + 2^0 y_0 + 2^1 y_1 + \dots + 2^a y_a = 0 \tag{3.2}$$

It can be shown that for any solution $(x_1,\dots,x_n)$ of (3.1) there exist $(y_o,\ldots,y_a) \in L_2^{a+1}$ so that $(x_1,\ldots,x_n\ ;\ y_o,\ldots,y_n)$ would be a solution of (3.1), and converesely, for any solution $(x_1,\ldots,x_n\ ;\ y_o^c,\ldots,y_a)$ of (3.2), $(x_1,\ldots,x_n)$ is a solution of (3.1). Taking into account the fact that **any** PBIN may be brought to the form (3.1), (because $f < 0$ means $f + 1 \leq 0$; $f \geq 0$ means $-f \leq 0$, $f > 0$ means $-f - 1 \leq 0$) we see that we may confine ourselves to the study of PBET.

<u>A system</u>

$$f_i(x_1,\ldots,x_n) = 0 \qquad (i=1,\ldots,m) \tag{3.3}$$

of PBET is obviously equivalent with the single PBET

$$\prod_{i=1}^{m} \left[f_i(x_1,\ldots,x_n) + 1 \right] = 1 \tag{3.4}$$

In conclusion, for studying systems pf PBET and PBIN, we may confine ourselves to the study of a single PBET.

Let us consider a PBET

$$f(x_1,\ldots,x_n) = 0. \tag{3.1}$$

A set of PBF :

$$x_i = x_i(u_1,\ldots,u_m) \qquad (i=1,\ldots,n) \tag{3.5}$$

is called the <u>general solution</u> of (3.1) if for any $(u_1,\ldots,u_m) \in L_2^m$ $(x_1,\ldots,x_n)$ is in L_2^n and is a solution of (3.1), and conversely, for any solution $(x_1^o,\ldots,x_n^o)$ of (3.1) there exist values $(u_1^o,\ldots,u_m^o) \in L_2^m$ so that

$$x_i^o = x_i(u_1^o,\ldots,u_n^o) \qquad (i=1,\ldots,n) \tag{3.6}$$

For any integer k let us define its reduced k' by

$$k' = \begin{cases} 1 & \text{if } k \neq 0 \\ 0 & \text{if } k = 0 \end{cases}$$

Analogously the reduced of a PBF $f(x_1,\ldots,x_n)$ is defined by

$$f'(x_1,\ldots,x_n)=(f(x_1,\ldots,x_n))' = \begin{cases} 1 & \text{if } f(x_1,\ldots,x_n) \neq 0 \\ 0 & \text{if } f(x_1,\ldots,x_n) = 0 \end{cases}$$

We have the following

THEOREM 3 (/59/). _If $(x_1^o,\ldots,x_n^o)$ is any particular solution of the pseudo-Boolean equation (3.1), then the general solution of it will be_

$$x_i = x_i^o \cdot f'(p_1,\ldots,p_n)+p_i \cdot \overline{f'(p_1,\ldots,p_n)} \qquad (3.7)$$

$$(i=1,\ldots,n)$$

where $p_1,\ldots,p_n$ are arbitrary parameters in L_2.

In /59/ we have also indicated a way to construct a particular solution of (3.1).

There is also another possibility of solving PBET (/6o/), and this extends the method of succesive eliminations for solving Boolean equations, described by S. Rudenau in /1o8/.

Let us write (3.1) in the form (putting f_1 instead of f):

$$f_1 = \varphi_1(x_2,\ldots,x_n).x_1 + \psi_1(x_2,\ldots,x_n)\overline{x}_1 = 0,$$

where φ_1 and ψ_1 are PBF of $x_2,\ldots,x_n$.

Let us further put

$$f_2 = \varphi_1(x_2,\ldots,x_n).\psi_1(x_2,\ldots,x_n)$$

and consider the equation

$$f_2 = \varphi_2(x_3,\ldots,x_n)x_2 + \psi_2(x_2,\ldots,x_n)\overline{x}_n = 0.$$

Analogously we consider

$$f_3 = \varphi_2\psi_2 = \varphi_3(x_4,\ldots,x_n)x_3 + \psi_3(x_4,\ldots,x_n)\overline{x}_3 = 0$$

. . . .

$$f_k = \varphi_{k-1}\varphi_{k-1} = \varphi_k(x_{k+1},\ldots,x_n)x_k + \psi_k(x_{k+1},\ldots,x_n)\overline{x}_k = 0$$

. . . .

$$f_n = \varphi_{n-1}\psi_{n-1} = \varphi_n x_n + \psi_n \overline{x}_n = 0$$

where φ_n and ψ_n are constants.

Finally, for any $(p_k,\ldots,p_n) \in L_2^{n-k+1}$ we put

$$X_k(p_k,\dots,p_n) = S_k(p_{k+1},\dots,p_n)\cdot p_k + T_k(p_{k+1},\dots,p_n)\cdot \bar{p}_k \qquad (3.8.k)$$

$(k=1,\dots,n-1)$

where

$$S_k(p_{k+1},\dots,p_n) = \overline{\varphi'_k\left[X_{k+1}(p_{k+1},\dots,p_n),\dots,X_n(p_n)\right]}$$

$$T_k(p_{k+1},\dots,p_n) = \psi'_k\left[X_{k+1}(p_{k+1},\dots,p_n),\dots,X_n(p_n)\right];$$

while

$$X_n(p_n) = \overline{\varphi'_n}p_n + \psi'_n\bar{p}_n \qquad (3.8.n)$$

It may be proved (/6o/) that

THEOREM 4 <u>If the equation (3.1) is solvable then the system</u>

$$x_k = X_k(p_k,\dots,p_n) \qquad (k=1,\dots,n) \qquad (3.9)$$

<u>defined by</u> (3.8.k), (3.8.n) <u>is a general solution of it</u>.

Both theorems 3 and 4 require the knowledge of the reduced of a PBF. Let us consider for this sake a PBF, $W(x_1,\dots,x_n)$ and write it in the form

$$\sum c_{\lambda\mu\dots\pi}\, x_1^{\lambda}\, x_m^{\mu}\dots x_p^{\pi} - \sum c_{\varrho\sigma\dots\tau}\, x_r^{\varrho}\, x_s^{\sigma}\dots x_t^{\tau} \qquad (3.1o)$$

where all the coefficients $c_{\alpha\beta\dots\gamma}$ are positive.

Using the bynary development of the coefficients

$$c_{\alpha\beta\dots\gamma} = 2^0 d^0_{\alpha\beta\dots\gamma} + 2^1 d^1_{\alpha\beta\dots\gamma} + \dots + 2^q d^q_{\alpha\beta\dots\gamma} \qquad (3.11)$$

we see that (3.1o) becomes

$$\sum_j 2^j \sum_{\lambda\mu\dots\pi} d^j_{\lambda\mu\dots\pi}\, x_1^{\lambda}\, x_m^{\mu}\dots x_p^{\pi} - \sum_j 2^j \sum_{\varrho\sigma\dots\tau} d^j_{\varrho\sigma\dots\tau}\, x_r^{\varrho}\, x_s^{\sigma}\dots x_t^{\tau} \qquad (3.12)$$

We can now easily determine (using for instance the formulas given by R. Fortet /37/, or by P. Camion /19/ the bynary developments of the two terms of (3-12);

$$\sum_{j=1}^{h} 2^j U_j\quad (x_1,\dots,x_n) - \sum_{j=1}^{k} 2^j V_j(x_1,\dots,x_n) \qquad (3\text{-}13)$$

If $h \neq k$, for instance if $h > k$, let us introduce $U_{k+1} = \dots = U_h = 0$.

It is proved in /61/ that

THEOREM 5

$$W'(x_1,\dots,x_n) = \prod_{j=1}^{h}\left[U_j(x_1,\dots,x_n)\Delta V_j(x_1,\dots,x_n)\Delta 1\right]\Delta 1. \quad (3.14)$$

§ 2 Using Congruences Modulo 2

P. Camion has shown in /19/ that for any PBET or PBIN, their exist a system of congruences in GF(2) with the same unknowns, so that any solution of the first is a solution of the second system, and conversely, any solution of the second system is a solution of the first. But, instead of studying a system of congruences

$$R_i(x_1,\dots,x_n)\equiv 0 \pmod 2 \qquad (i=1,\dots,m) \quad (3.15)$$

we may study the single congruence

$$R(x_1,\dots,x_n)\equiv 0 \pmod 2 \quad (3.16)$$

where

$$R(x_1,\dots,x_n)\equiv \prod_{j-1}^{m}\left[R_1(x_1,\dots,x_n)\Delta 1\right]\Delta 1 \quad (3.17)$$

In /56/ we have proved

THEOREM 6 Given any particular solution $(x_1^o,\dots,x_n^o)$ of (3.16), its general solution is

$$x_j \equiv (x_j^o\Delta p_j)\, R(p_1,\dots,p_n)\Delta p_j. \pmod 2 \quad (3.18)$$

where $p_1,\dots,p_n$ are arbitrary parameters in GF(2).

Here, the general solution of a congruence modulo 2 is analogously defined as that of a PBF (Chapter 3, § 1).

As any Galoisian function has a polynomial expression (Theorem I, Chapter II, § 1 of /96/), we may out (writting R_1 instead of R)

$$R_1(x_1,\dots,x_n)=x_1R_1^1(x_2,\dots,x_n)\Delta R_1^2(x_2,\dots,x_n)\equiv 0 \pmod 2 \quad (3.19)$$

and let

$$R_2\equiv R_1^1(x_2,\dots,x_n).\ R_1^2(x_2,\dots,x_n)\Delta R_1^2(x_2,\dots,x_n) \pmod 2$$

We have the congruence

$$R_2(x_2,\dots,x_n) = x_2R_2^1(x_2,\dots,x_n)\Delta R_2^2(x_2,\dots,x_n)\equiv 0 \pmod 2 \qquad (3.19.2)$$

and analogously

$$R_j = x_jR_j^1(x_2,\dots,x_n)\Delta R_j^2(x_2,\dots,x_n)\equiv 0 \pmod 2 \qquad (3.19.j)$$

. . . .

$$R_n = x_n\ R_n^1\Delta R_n^2 \equiv 0 \pmod 2 \qquad (3.19.n)$$

We have proved (/56/)

THEOREM 7 <u>A necessary and sufficient condition of solvability for the congruence</u> (3.16) <u>is that</u>

$$R_n^1\ R_n^2\Delta R_n^2\equiv 0 \pmod 2 \qquad (3.20)$$

<u>and its general solution is</u>

$$x_j(p_1,\dots,p_n)\equiv\left[R_j^1(p_{j+1},\dots,p_n)\Delta 1\right]\ p_j\Delta R_j^2(p_{j+1},\dots,p_n) \pmod 2 \qquad (3.21)$$

<u>for</u> $j=1,\dots,n$, <u>where</u> $p_1,\dots,p_n$ are arbitrary parameters in GF(2).

§ 3 Using Boolean Equations

This procedure is proposed in /64/. R. Fortet has shown in /37/ that any PBET or PBIN is equivalent with a set of Boolean equations. But a set of Boolean equations is equivalent with a single Boolean equation. To obtain the general solution of it we may use the methods of Johnson /77/, Löwenheim /86/, Rudeanu /108/, /109/, /110/, etc.

Chapter 4

Pseudo-Boolean Programming

In this chapter we shall examine the possibility of applying the methods given in chapters 2 and 3 for the solution of the following problem:

Determine

$$(x_1,\dots,x_n) \in L_2^n \tag{4.1}$$

satisfying the PBR

$$f_i(x_1,\dots,x_n)\ R_i 0 \qquad (i=1,\dots,m) \tag{4.2}$$

(R_i being the relations $=, \leq, \geq, <, >$) and minimizing the PBF

$$f_o(x_1,\dots,x_n) \tag{4.3}$$

For a detailed presentation see /55/, /56/, /62/, /63/, /64/.

§ 1 The general Procedure

The general procedure for solving the above problem consists in two phases:

Phase 1 Solve the system (4.2) with one of the methods described in Chapter 3. Let the general solution of (4.2) be

$$x_j = x_j(p_1,\dots,p_n) \qquad (j=1,\dots,n) \tag{4.4}$$

where $p_1,\dots,p_n$ are arbitrary parameters in L_2.

Phase II Introduce (4.4) in (4.3); let us denote

$$F(p_1,\dots,p_n) = f_o\left[x_1(p_1,\dots,p_n),\dots,x_n(p_1,\dots,p_n)\right] \tag{4.5}$$

Minimizing now the unconstrained PBF $F(p_1,\dots,p_n)$ with the method given in chapter 2, we find

$$p_j = p_j(u_1,\dots,u_n) \qquad (j=1,\dots,n) \tag{4.6}$$

where $u_1,\dots,u_n$ are arbitrary parameters in L_2.

Introducing (4.6) in (4.4) and this in (4.3), we find

$$x_j = X_j(u_1,\dots,u_n) \tag{4.7}$$

and

$$\min f_o = f_o\left[X_1(u_1,\dots,u_n),\dots,X_n(u_1,\dots,u_n)\right]; \tag{4.8}$$

for any value of $(u_1,\ldots,u_n)\in L_2^n$ we obtain a solution of the problem and every solution of it may be obtained in this way.

§ 2 Problems with Non-Negative Restraints

In many a problem of pseudo-Boolean programming the function f_i of (4.2) have the special property that for any $(x_1,\ldots,x_n)\in L_2^n$,

$$f_i(x_1,\ldots,x_n)\geqslant 0. \qquad (4.9)$$

In this case denoting with D and E sums of the positive and negative coefficients of a polynomial expression of $f_o(x_1,\ldots x_n)$, we have

THEOREM 8 The solution of the problem (4.1), (4.2), (4.3) when condition (4.9) takes place may be obtained by minimizing (with the method given in chapter 2) the unconstrained PBF :

$$f_o(x_1,\ldots,x_n)+ (D-E+1)\sum_{i=1}^{m} f_i(x_1,\ldots,x_n). \qquad (4.1o)$$

In Part II we shall have numerous applications of theorem 8.

Chapter 5

Discrete Polynomial-Logical Programming

The importance of solving problems of mathematical programming in integers was repeatedly pointed out by G.B.Dantzig (/22/,/23/,/24/) and the procedures of R. Gomory /43/,/44/, E.M.L. Beale /14/, A.Ben-Israel and A.Charnes /17/, A.H.Land and A.G.Doig /85/, E.Balas /2/,/3/, and of others solve this problem for the linear case. H.P.Künzi and W.Oettli's method /83/, solves the problem in case when the objective function is quadratic and convex, the constraints being linear. On the other hand L. Németi and F. Rado /98/, /1o3/ have studied problems of linear programming with logical conditions.

In this chapter we shall show how to reduce to a problem of pseudo-Boolean programming, and hence how to solve it with the method of chapter 2, problems of the following type:

Let Y denote any finite subset of Z,

Determine the elements

$$y_1,\ldots,y_n \text{ in } Y \tag{5.1}$$

and

$$z_1,\ldots,z_k \text{ in } Z \tag{5.2}$$

minimizing the polynomial

$$p_o(y_1,\ldots,y_h\ ;\ z_1,\ldots,z_k) \tag{5.3}$$

with integer coefficients, and satisfying

$$\underline{M}_q \leqslant z_q \leqslant \overline{M}_q \qquad (q=1,\ldots,k) \tag{5.4}$$

$$p_i(y_1,\ldots,y_h\ ;\ z_1,\ldots,z_k)\ R_i 0 \tag{5.5}$$

$$\text{"if } p_j'(y_1,\ldots,y_h;z_1,\ldots,z_k)R'0 \text{ then } p_j''(y_1,\ldots,y_h;z_1,\ldots,z_k)R''0\text{ "} \tag{5.5}$$

"at least one of the relations

$$\left.\begin{array}{l} p_j'(y_1,\ldots,y_h\ ;\ z_1,\ldots,z_k)\ R'0 \\ p_j''(y_1,\ldots,y_h\ ;\ z_1,\ldots,z_k)\ R''0 \end{array}\right\}(x) \tag{5.6}$$

holds"

"exactly one of the relations (x) holds". (5.7)

where p' and p'' are polynomials with integer coefficients, R_i, R', R'' are the relations $=, \leqslant, \geqslant, <, >$, and where $\underline{M}_q$, $\overline{M}_q$ are given integers.

§ 1 Integer Polynomial Programming

The problem of integer polynomial programming is a special case of the above one, in which $Y = \emptyset$, and in which conditions (5.5), (5.6), (5.7) are absent.

Let us put

$$\overline{z}_q = \underline{M}_q + z'_q \tag{5.8}$$

$$\overline{M}_q - \underline{M}_q = m_q \tag{5.9}$$

and let

$$z'_q = 2^0 z'_{oq} + 2^1 z'_{1q} + \ldots + 2^{s(q)} z'_{s(q),q} \tag{5.1o}$$

be the bynary development of the non-negative integers z'_q. Introducing (5.1o) in (5.8) and this new expression of z_q in (5.3), (5.4), (5.5), the initial problem is reduced to that of minimizing a PBF subject to some PBR, i.e. the problem becomes one of pseudo-Boolean programming.

§ 2 Discrete Polynomial Programming

The problem of discrete polynomial programming is a special case of that described at the beginning of this chapter, in which conditions (5.6), (5.7), (5.8) are absent.

It may be easily converted into one of integer polynomial programming by attaching to any element of the finite set Y a bivalent variable: if this bivalent variable proves to be equal to one in the final solution of the problem, then the corresponding y is in the optimal solution, otherwise not.

§ 3 Logical Conditions

We have remarked (Chapter 3, § 1) that any pseudo-Boolean inequality (p',p") may be converted into a pseudo-Boolean equation (p',p"). Therefore in (5.5), (5.6), we may consider all the R'-s and R"-s as being equalities

In this case introducing new bivalent variables α, β, we see that:

- Condition (5.5) is equivalent with

$$[p'_j]' \cup [\overline{p''_j}]' = 1 \tag{5.5'}$$

(where $\left[p_j'\right]'$ stands for the reduced of p_j')

- Condition (5.6) is equivalent with

$$\left.\begin{aligned} \alpha p_j' + \beta p_j'' &= 0 \\ \alpha \cup \beta &= 1 \end{aligned}\right\} \qquad (5.6')$$

- Condition (5.7) is equivalent with

$$\begin{aligned} \left[p_j' + p_j''\right]' &= 1 \\ \alpha p_j' + \beta p_j'' &= 0 \\ \alpha + \beta &= 1 \end{aligned} \qquad (5.7')$$

That means that conditions (5.5),(5.6),(5.7) may also be brought to the form of PBR, thus completing our discussion.

PART II

Chapter 6

Application to the Theory of Graphs

The aim of the present chapter is to apply the method of pseudo-Boolean programming to the determination of the basic numbers (chromatic number, number of internal stability, number of external stability) and of the kernels of a finite graph. For proofs, details and examples see /69/. The terminology is that of /18 bis/.

Khaled Maghout /89/ solves similar problems of the theory of graphs with a Boolean apparatus, but the approach of that paper and of ours are of different types.

By a graph $G = (V, \rho)$ we shall mean a finite non-empty set $V = \{v_1, \ldots, v_n\}$ of elements called vertices, and a multivalued mapping of V into itself. An ordered pair (v_i, v_j) of elements of V is called an edge if $v_j \in \rho v_i$. We shall suppose that for any i, $v_i \notin \rho v_i$.

We define for any graph $G = (V, \varrho)$ a $(n \times n)$ matrix $C_G = ((c_{ij}))$ by setting

$$c_{ij} = \begin{cases} 1 & \text{if } v_j \in \varrho v_i \\ 0 & \text{if } v_j \notin \varrho v_i \end{cases}$$

For any set $M \subseteq V$, the characteristic fonction $\chi_M(V)$ is defined by:

$$x_i^M = \chi_M(v_i) = \begin{cases} 1 & \text{if } v_i \in M \\ 0 & \text{if } v_i \notin M \end{cases}$$

Thus, any set $M \subseteq V$ is characterized with an n-tuple $(x_1^M, \ldots, x_n^M)$ of zeroes and ones.

By $|A|$ we mean the power of the set A.

§ 1 Determination of the Chromatic Number

A set $R \subseteq V$ is called interiorly stable, if $\varrho R \cap R = \emptyset$, i.e.

$$v_i \in R,\ c_{ij} = 1 \Longrightarrow v_j \notin R. \tag{6.1}$$

Given a finite graph $G = (V, \varrho)$, by a chromatic decomposition of it we mean a family of interiorly stable disjoint subsets $M_1, \ldots, M_k$ of V so that $\bigcup_{h=1}^{k} M_h = V$. A chromatic decomposition with the smalles number $\gamma(G)$ of subsets is called a minimal chromatic decomposition and $\gamma(G)$ is the chromatic number of the graph.

An interiorly stable subset M of $G = (V, \varrho)$ is called superior if for any interiorly stable subset N of G, $M \subseteq N$ implies $M = N$. We have (/69/):

THEOREM 9 *The chromatic number $\gamma(G)$ of the graph $G = (V, \varrho)$ is equal to the minimal number of superior interiorly stables subsets covering V.*

Let us denote with $\mathcal{M} = \{M_1, \ldots, M_q\}$ the family of all superior interiorly stable subsets of V ; they may be determined by minimizing the PBF

$$W = \sum_{i=1}^{n} \sum_{j=1}^{n} c_{ij} x_i x_j \qquad (6.2)$$

and putting

$$M_u = \left\{ v_{j_1}, \ldots, v_{j_{q_u}} \right\} \qquad (6.3)$$

if and only if W has a minimum at

$$x_j = \begin{cases} 1 & \text{if } j \in \{j_1, \ldots, j_{q_u}\} \\ 0 & \text{if } j \notin \{j_1, \ldots, j_{q_u}\} \end{cases} \qquad (6.4)$$

Let $d_{ij} = 1$ if v_i belongs to the superior interbrly stable set M_j, and $d_{ij} = 0$ contrarly. We have:

THEOREM 10 <u>The chromatic number $\gamma(G)$ of the graph $G = (V, \rho)$ is equal to the minimum of the pseudo-Boolean function</u>

$$K = (q+1) \sum_{i=1}^{n} \prod_{j=1}^{q} (1 - d_{ij} x_j) + \sum_{h=1}^{q} x_h \; ; \qquad (6.5)$$

<u>if $K_{min} = K(x_1^o, \ldots, x_n^o)$, then a minimal chromatic decompositions $P_1, \ldots, P_\gamma$ of G may be obtained from the family of all superior interiorly stable subsets M_j of V for which $x_j^o = 1$ by putting:</u>

$$\left. \begin{aligned} P_1 &= M_1 \\ P_2 &= M_2 - M_1 \\ &\ldots\ldots\ldots \\ P_i &= M_i - \bigcup_{j=1}^{i-1} M_j \\ &\ldots\ldots\ldots \\ P_\gamma &= M - \bigcup_{j=1}^{\gamma-1} M_j \end{aligned} \right\} \qquad (6.6)$$

§ 2 Determination of the Number of Internal Stability

Let us denote with $\mathcal{R}$ the family of all interiorly stable sets of a graph G ; the number of internal stability of G is

$$\alpha(G) = \max_{R \in \mathcal{R}} |R| \qquad (6.7)$$

We have

THEOREM 11 <u>For any</u> $(x_1^o,\dots,x_n^o) \in L_2^n$ <u>minimizing the pseudo-Boolean function</u>

$$E = (n+1)\sum_{i=1}^{n} \sum_{j=1}^{n} c_{ij}x_i x_j - \sum_{k=1}^{n} x_k \qquad (6.8)$$

<u>and for</u>

$$R^o = \left\{ v_i \;\middle|\; v_i \in V,\ x_i^o = 1 \right\} \qquad (6.9)$$

we have

$$\alpha(G) = \sum_{i=1}^{n} x_i^o = \left| R^o \right| \qquad (6.10)$$

<u>and any maximal interiorly stable set may be obtained in this way</u>.

§ 3 Determination of the Number of External Stability

A set $S \subseteq V$ is called externaly stable if for any $s \in S$, $\rho s \cap S \neq \emptyset$, i.e.

$$v_i \notin S \Longrightarrow (\exists)\ v_j \in S,\ c_{ij} = 1 \qquad (6.11)$$

Let $\mathcal{S}$ denote the family of all externaly stable sets of a graph G; the <u>number of external stability</u> of G is

$$\beta(G) = \min_{S \in \mathcal{S}} |S| \qquad (6.12)$$

We have

THEOREM 12 <u>For any</u> $(x_1^o,\dots,x_n^o)$ <u>minimizing the pseudo-Boolean function</u>

$$H = (n+1) \sum_{i=1}^{n} \prod_{j=1}^{n} \left[1 - (c_{ij}+\delta_i^j)x_j \right] + \sum_{k=1}^{n} x_k \qquad (6.13)$$

and for

$$S^{o} = \left\{ v_i \middle| v_i \in V,\ x_i^{o} = 1 \right\} \qquad (6.14)$$

we have

$$\beta(G) = \sum_{i=1}^{n} x_i^{o} = \left| S^{o} \right| \qquad (6.15)$$

and any minimal exteriorly stable set may be obtained in this way (here δ_i^j is the Kronecker symbol).

§ 4 Determination of the Kernels

A set $T \subseteq V$ being simultaneously interiorly and exteriorly stable, is called a kernel of the graph G.

We have

THEOREM 13 The graph $G = (V, \mathcal{S})$ posesses a kernel if and only if the minimum of the pseudo-Boolean function

$$J = \sum_{i=1}^{n} (1-x_i) \prod_{j=1}^{n} (1-c_{ij}x_j) + \sum_{i=1}^{n} \sum_{j=1}^{n} c_{ij}x_i x_j \qquad (6.16)$$

is equal to zero. In this case, if $J_{min} = J(x_1^{o}, \dots, x_n^{o}) = 0$, then

$$T = \left\{ v_i \middle| v_i \in V,\ x_i^{o} = 1 \right\} \qquad (6.17)$$

is a kernel, and any kernel may be obtained in this way.

Chapter 7

Applications to the Theory of Flows in Networks

The aim of the present chapter is to apply the method of pseudo-Boolean programming to the determination of the minimal cut and of the value of the maximal flow through a network without (§ 1) or with (§ 2) given lower bounds on the arc flows as well as the solution of some feasibility problems in networks (§ 3). Throughout this chapter we shall use the terminology of

L.R. Ford and D.R. Fulkerson (/34/). Systematic expositions of the basic concepts of the theory of flows in networks may be also found in /4o/,/42/,/112/. For proofs, details and examples regarding the material of this chapter, see /57/.

By a network $\mathbf{N} = (W, \sigma)$ we mean a finite graph in which the node w_1 (called the source) has the property that

$$w_1 \notin \sigma w_i, \qquad (i=1,2,\ldots,n) \qquad (7.1)$$

in which the node w_n (called the sink) has the property that

$$w_i \notin \sigma w_n, \qquad (i=1,2,\ldots,n) \qquad (7.2)$$

and for which we are given a n x n matrix $K = ((k_{ij}))$ of non-negative integers (called arc capacities) satisfying

$$(c_{ij} = 0) \Longrightarrow (k_{ij} = 0) \qquad (7.3)$$

§ 1 Minimal Cuts in a Network

A flow of value L from the source w_1 to the sink w_n is a n x n matrix $F = ((f_{ij}))$ of non-negative elements called arc-flows satisfying the following conditions:

$$\sum_{j=1}^{n} c_{ij} f_{ij} - \sum_{j=1}^{n} c_{ji} f_{ji} = \begin{cases} L & \text{if } i = 1 \\ 0 & \text{if } i \neq 1,n \\ -L & \text{if } i = n \end{cases} \qquad (7.4)$$

$$f_{ij} \leq k_{ij} \qquad (7.5)$$

The maximal flow problems is that of finding those f_{ij} for which L would be maximal, the conditions (7.4) and (7.5) being fulfilled.

If $U \subseteq W$ we shall put $U' = W - U$ and $I_U = \{i \mid w_i \in U\}$. For any integer-valued function defined on the arcs of G, and having the value g_{ij} on the arc (w_i, w_j), we put

$$g(U,U') = \sum_{i \in I_u} \sum_{j \in I_{U'}} g_{ij} \qquad (7.6)$$

By a <u>cut</u> $\mathcal{C}$ in G, separating w_1 and w_n we mean a set $Q \subseteq W$ so that $w_1 \in Q$, $w_n \in Q'$. The capacity of the cut is $k(Q,Q')$.

The minimal cut problem is that of determining a cut of minimal capacity.

A cut Q separating w_1 and w_n may be characterized by a vector $(1,q_2,\ldots,q_{n-1},0) \in L_2^n$ where q_i is equal to 1 if $i \in I_Q$ and to 0 if $i \notin I_Q$.

For the computation of the minimal cut we have given /57/ the following

THEOREM 14 *If* $(1,q_2^o,\ldots,q_{n-1}^o,0)$ *gives a minimum of the pseudo-Boolean function*

$$k_{1n} + \sum_{j=2}^{n-1} k_{1j}\bar{q}_j + \sum_{i=2}^{n-1} k_{in}q_i + \sum_{i=2}^{n-1} \sum_{j=2}^{n-1} k_{ij}q_i\bar{q}_j \qquad (7.7)$$

then

$$Q_o = \left\{ w_i \mid q_i^o = 1 \right\} \qquad (7.8)$$

is a minimal cut separating w_1 *and* w_2*, and conversely, any minimal cut separating* w_1 *and* w_n *may be obtained in this way*.

According to the basic theorem of Ford and Fulkerson (/32/,/33/) we may compute the value of the maximal flow with the formula

$$L_{max} = k(Q_o,Q_o'). \qquad (7.9)$$

§ 2 The Case of Lower Bounds on Flows

Let us suppose that we are given a n x n matrix $H = ((h_{ij}))$ of non-negative integers called <u>lower bounds</u>, satisfying

$$0 \leq h_{ij} \leq k_{ij} \tag{7.10}$$

and we are seeking the maximal L for which f_{ij} satisfy (7.4) and

$$h_{ij} \leq f_{ij} \leq k_{ij} \tag{7.11}$$

Using a result of Ford and Fulkerson /34/, we have proved /57/:

THEOREM 15 <u>If there is a matrix</u> $((f_{ij}))$ <u>satisfying</u> (7.4) <u>and</u> (7.11) <u>for some L, then the maximal value of L subject to these constraints is equal to the minimum of the pseudo-Boolean function</u>:

$$k_{1n} - h_{n1} + \sum_{j=2}^{n-1} (k_{1j}\bar{q}_j - k_{nj}q_j) + \sum_{i=2}^{n-1} (k_{in}q_i - h_{i1}\bar{q}_{i1}) +$$

$$+ \sum_{i=2}^{n-1} \sum_{j=2}^{n-1} (k_{ij}q_i\bar{q}_j - h_i\bar{q}_i q_j) \tag{7.12}$$

§ 3 <u>Some Feasibility Theorems</u>

We have proved in /57/, that

- Gale's supply-demand theorem /41/,
- Fulkerson's symmetric supply-demand theorem /39/,
- Hoffman's circulation theorems /51/

may all be given a unified treatment using pseudo-Boolean programming. As a result, the feasibility conditions of the above theorems, are expressed as non-negativitty conditions of certain pseudo-Boolean functions.

Chapter 8

Applications to the Transportation Problem

Egerváry's method for solving transportation problems /29/, (but see also /81/, /82/), involves an insolved step[8] ;

this step may be handled with the method of pseudo-Boolean programming. For proofs, details and examples see /73/, /74/, /75/.

The problem is the following: A matrix and a set of "distinguished" elements of it (in fact its zeroes) are given. To each row i and to each column j, a positive real number a_i and b_j, respectively is associated. The problem consists in finding a system

$$S = \left\{ i_1, \ldots, i_p \; ; \; j_1, \ldots, j_q \right\} \qquad (8.1)$$

of rows and columns, covering all the distinghuished elements of the matrix (i.e. any distinghuished element of it belongs to a row or/and a column of S) and so that the sum

$$F(S) = \sum_{h=1}^{p} a_{i_h} + \sum_{k=1}^{q} b_{j_k} \qquad (8.2)$$

be minimized.

This problem leads us to that of finding the minimum of the function

$$F(y_1, \ldots, y_m \; ; \; z_1, \ldots, z_n) = \sum_{i=1}^{m} a_i y_i + \sum_{j=1}^{n} b_j z_j \qquad (8.3)$$

where the Boolean variables y_i, z_j are subject to the restrictions

$$y_i \cup z_j = 1 \quad \text{if } i \in S_j, \qquad (8.4)$$

S_j being the set of distinguished elements of column j. This problem of pseudo-Boolean programming, is reduced to the following problem of minimization of an unconstrained pseudo-Boolean function:

Find $(x_1, \ldots, x_n) \in L_2^n$ minimizing

$$G(x_1, \ldots, x_n) = \sum_{i=1}^{n} a_i x_i - \sum_{j=1}^{n} b_j \prod_{i \in S_j} x_i \qquad (8.5)$$

---

8) A method for solving it is given by B. Krekó in the Appendix of /84/; the problem was also studied by I. Kolumban /78/.

After solving this problem with the method given in chapter 2, putting

$$y_i^o = x_i^o \quad (i=1,\ldots,n) \tag{8.6}$$

$$z^o = \prod_{i \in S_j} x_i^o \quad (j=1,\ldots,n) \tag{8.7}$$

$$F(y_1^o,\ldots,y_m^o \; ; \; z_1^o,\ldots,z_n^o) = G(x_1^o,\ldots,x_n^o) \tag{8.8}$$

we obtain the optimal solution of the initial problem.

The computational time required by the above method in solving concrete transportation problems, turned out to be sensibly reduced, compared to that required by other methods.

Chapter 9

Applications to Switching Algebra

In this chapter we shall describe a method of determining all absolute minima of a Boolean function. This problem arises frequently in switching algebra where it is interpreted as the problem of determining a switching circuit satisfying a given program of functioning and containing for instance the smallest number of relays (this criterion of economcity may be present also in other, slightly different, forms). Numerous methods are available for determinating relative minima of a Boolean function (e.g.[9]) MacCluskey /88/ and Quine /1o1/,/1o2/; for a systematic presentation of them see /2o/,/27/,/31/,/96/. As to our knowledge at present this method is the single one which obtains (i) absolute (and not relative) minima; and (ii) all the absolute minima of a Boolean function. For proofs, details and examples see /72/.

The problem may be described as follows: L_2^n is partitioned into three disjoint subsets: N_f, Z_f and D_f. The problem is

9) See also /9o/, /1o5/, /1o6/, /1o7/, /114/, /115/.

to find the expression of a Boolean function $f(x_1,\dots,x_n)$ so that

$$f(x_1,\dots,x_n) = \begin{cases} 1 & \text{if } (x_1,\dots,x_n)\in N_f \\ 0 & \text{if } (x_1,\dots,x_n)\in Z_f \\ \text{arbitrary} & \text{if } (x_1,\dots,x_n)\in D_f \end{cases} \qquad (9.1)$$

and so that the number of letters[1o] (with or without bars) should be minimal. An expression of f satisfying the above conditions is an absolute minimum of f. A relative minimum of f is an expression satisfying (9.1) and so that deleting any letter or term of it the resulting expression ceases to fulfil (9.1). A Boolean function $p(x_1,\dots,x_n)$ is called an implicant of f if

$$(p(x_1,\dots,x_n) = 1) \Longrightarrow (f(x_1,\dots,x_n) = 1) \qquad (9.2)$$

An implicant is called prime if deleting any letter or term of it, it ceases to be an implicant.

Let $p_1(x_1,\dots,x_n),\dots,p_m(x_1,\dots,x_n)$ denote the prime implicants of a Boolean function $f'(x_1,\dots,x_n)$ equal to one on $N_f \cup D_f$ and equal to zero on Z_f; $p_1,\dots,p_m$ may be easily determined with a method of MacCluskey /88/.

Let us denote with $X_1, X_2,\dots,X_k$ the elements of N_f and let us put

$$a_{ij} = \begin{cases} 1 \text{ if } & X_i \text{ is an implicant of } p_j \\ 0 \text{ if } & X_i \text{ is not an implicant of } p_j \end{cases} \qquad (9.3)$$

Denoting with π_j the number of letters contained in p_j, and putting

$$\pi = \sum_{j-1}^{m} \pi_j + 1$$

we have (/72/):

1o) In fact the method permits also the choice of any other criterion of minimization, which fulfiles certains axioms (see /72/).

THEOREM 16 *If* $(y_1^o,\dots,y_m^o) \in L_2^m$ *is a minimum of the pseudo-Boolean function*

$$\sum_{j=1}^{m} \pi_j y_j + \pi \sum_{j=1}^{k} \prod_{j=1}^{m} (1 - a_{ij} y_j) \tag{9.4}$$

then the absolut minimum of f is

$$\bigcup_{j=1}^{m} y_j^o p_j(x_1,\dots,x_n), \tag{9.5}$$

and any absolut minimum of **f** *is of this type.*

Chapter 1o

Minimal decomposition of finite partially ordered sets in chains

The aim of this chapter is to show that the problem of finding the number N of chains in a minimal decomposition of a finite partially ordered set S, as well as that of actually determining those chains, may be reduced to one of minimizing a pseudo-Boolean function. The results of Dilworth /28/, Dantzig and Hoffman /25/, and of Fulkerson /38/ on the number of chains in a minimal decomposition of S, offer various means of translating this problem into the pseudo-Boolean language. For proofs and details see /58/.

Let us consider a finite partially ordered set $S = \{s_1,\dots,s_n\}$ with its order relation denoted by "$\prec$". Let us form a n x n matrix $B = ((b_{ij}))$, by putting

$$b_{ij} = \begin{cases} 1 & \text{if } s_i \prec s_j \\ 0 & \text{if } s_i \not\prec s_j \end{cases} \tag{1o.1}$$

We shall suppose that $b_{ii} = 0$, for any i.

A <u>chain</u> is a subset $C = \{s_{i_1}, \dots, s_{i_k}\}$ of S, so that for any $s_{i_m}, s_{i_h} \in C$, either $s_{i_m} \prec s_{i_h}$ or $s_{i_h} \prec s_{i_m}$. An <u>anti-chain</u> is a subset $D = \{s_{j_1}, \dots, s_{j_r}\}$ of S, so that for any $s_{j_u}, s_{j_v} \in D$, neither $s_{j_u} \prec s_{j_v}$, nor $s_{j_v} \prec s_{j_u}$. A decomposition of S is a family $\{C_1, \dots, C_N\}$ of disjunct chains, the union of which is S. A decomposition $\{C_1, \dots, C_N\}$ is called minimal, if for any other decomposition $\{C'_1, \dots, C'_{N'}\}$, $N \leq N'$.

Let us consider a finite graph $G = (V, A')$ and two subsets V' and V'' of V, so that

$$V' \cap V'' = \emptyset \qquad (10.2)$$

$$V' \cup V'' = V \qquad (10.3)$$

A (V', V'')-cut C of G is a subset of V, so that every arc joining a vertice of V' to one of V'' has at least one end in C. A minimal (V', V'')-cut is one having the smallest number of elements. A <u>(V', V'')-join</u> J of G is a subset of the set of arcs, each of them joining a vertice of V' to one of V'', and no two of them having a vertice in common. A maximal <u>(V', V'')-join</u> is one having the greatest number of elements.

§ 1 Maximal Anti-Chains

Dilworth /28/ has proved that the number N of chains in a minimal decomposition of a finite partially ordered set, is equal to the number M of elements contained in a maximal anti-chain of S. We have the following theorem for determining all the maximal anti-chains of S, /58/.

THEOREM 17 <u>Let $(d_1^o, \dots, d_n^o) \in L_2^n$ be a minimum of the pseudo-Boolean function</u>

$$F = - \sum_{h=1}^{n} d_h + (n+1) \sum_{i=1}^{n} \sum_{j=1}^{n} b_{ij} d_i d_j ; \qquad (10.4)$$

then

$$D^{o} = \left\{ s_{i} \;\middle|\; d_{i}^{o} = 1, \quad i=1,\ldots,n \right\} \qquad (1o.5)$$

is a maximal anit-chain of S, and the number of its elements is

$$M = F(d_{1}^{o},\ldots,d_{n}^{o}) \qquad (1o.6)$$

Conversely, given a maximal anti-chain $D^{o} = \left\{ s_{i_1},\ldots,s_{i_m} \right\}$ of S, we have m=M, and it exists a minimum $(d_{1}^{o},\ldots,d_{n}^{o})$ of the pseudo-Boolean function F so that (1o.5) holds.

From Dilworth's theorem we have the following

COROLARY

$$N = F(d_{1}^{o},\ldots,d_{n}^{o}) \qquad (1o.7)$$

§ 2 The Dantzig-Hoffman Formula

Let T be an arbitrary subset of S and let us put

$$T^{*} = \left\{ s_{i} \in S \;\middle|\; \exists s_{j} \in T,\ s_{i} \succ s_{j} \right\} \qquad (1o.8)$$

G.B. Dantzig and A.J. Hoffman /25/ have proved that

$$N = \max_{T \subseteq S} \left\{ |T| - |T^{*}| \right\}. \qquad (1o.9)$$

Using this result we have proved in /58/ the following

THEOREM 18 If $(t_{1}^{o},\ldots,t_{n}^{o}) \in L_2^n$ gives a minimum of the pseudo-Boolean function

$$G = -\sum_{h=1}^{n} t_h - \sum_{j=1}^{n} \prod_{b_{ij}=1}^{n} \bar{t}_i + n \qquad (1o.1o)$$

then

$$N = -G(t_{1}^{o},\ldots,t_{n}^{o}) \qquad (1o.11)$$

§ 3 Cuts and Joins

Fulkerson /38/ has pointed out the strong relationship between the problem of finding the minimal decomposition of S and the problem of finding a minimal cut or a maximal join of a certain graph G. G is defined as consisting of 2n vertices $U = \{v_1,\ldots,v_n, v_{n+1},\ldots,v_{2n}\}$ and arcs defined from S by the rule: if $s_i \prec s_j$ then (v_i, v_{n+j}) is an arc of G, and these are all the arcs of G. Let $V' = \{v_1,\ldots,v_n\}$, $V'' = \{v_{n+1},\ldots,v_{2n}\}$. The cuts and joins below are relative to (V', V''). The results of Fulkerson based on a theorem of König (/79/, p.232) permit to construct a minimal decomposition of S if we know a maximal join of G, and to construct a maximal anti-chain of S, if we know a minimal cut of G.

We have (/58/):

THEOREM 19 *Let* $(c_1^o,\ldots,c_n^o\,;\,c_{n+1}^o,\ldots,c_{2n}^o) \in L_2^{2n}$ *be a minimum of the pseudo-Boolean function*

$$H = -\sum_{h=1}^{2n} \bar{c}_h + (2n+1)\sum_{i=1}^{2n}\sum_{j=1}^{2n} b_{ij}\bar{c}_i\bar{c}_j. \qquad (10.12)$$

Then

$$C^o = \{v_i \mid c_i^o = 1, \quad i=1,\ldots,2n\} \qquad (10.13)$$

is a minimal cut of G and the number of its elements is

$$D = H(c_1^o,\ldots,c_{2n}^o) \qquad (10.14)$$

Conversely, given a minimal cut $C^o = \{v_{i_1},\ldots,v_{i_d}\}$ *of G, we have* $d = D$, *and it exists a minimum* $(c_1^o,\ldots c_{2n}^o)$ *of the pseudo-Boolean function H, so that (10.13) holds.*

From Fulkerson's results we have the following

COROLARY

$$N = H(c_1^o,\ldots,c_{2n}^o) \qquad (10.15)$$

Appendix

In this appendix we shall briefly mention some of the research-topycs on pseudo-Boolean programming which are in progress at this time.

1) Pseudo-Boolean programming viewed as dynamic programming /15/ with bivalent variables /65/. Generalization to a dynamic programming with trivalent variables; generalization to a dynamic programming with n-valent variables.

2) Accelerating the computations related to the determination of <u>all</u> $(x_1, \ldots, x_n) \in L_2^n$ for which a given pseudo-Boolean function is negative (or for which it is zero) /66/.

3) Solving problems related to matchings of byparthite graphs (/18/ bis/, /47/, /48/, /8o/, /81/, /91/, /99/, /1o4/, etc) with pseudo-Boolean programming /67/. Generalization for arbitrary graphs. Determination of the maximal cliques (/3o/,/49/,/87/) and of the degree of unbalance (/1/,/3o/) of a graph.

4) Systems of distinct representatives (/18 bis/, /34/, /45/,/46/,/47/,/48/,/51/,/52/,/53/,/79/,/91/,/92/,/99/,/1oo/, /1o4/,/111/), treated with pseudo-Boolean programming /68/.

5) Minimization of the working function of a switching circuit in case of hasards /97/ with pseudo-Boolean programming /76/. Minimization in Shefferian algebras /93/, /54/.

6) Solving the problem of optimal assignment of numbers to vertices /5o/ which arises in coding theory, with pseudo-Boolean programming /26/.

*

Acknowledgements

are due to my colleagues Sergiu Rudeanu, Ivo Rosenberg and Aristide Deleanu, for their kind permission to use in this lecture some of our common results.

REFERENCES

/1/. ABELSON, R.P., ROSENBERG, N.J.: "Symbolic Psycho-Logic: A Model of Attitudinal Cognition", Behaviour Sci.,3,1958, pp.1-13.

/2/. BALAS, E.: "Un algorithme additif pour la résolution des programmes linéaires en variables bivalentes",C.R.Acad. Sci., Paris,258,1964,n.15.

/3/. BALAS, E.: "Extension de l'algorithme additif à la programmation en nombres entiers et à la programmation non linéaire", C.R.Acad.Sci., Paris,t.258,1964,n.21.

/4/. BALAS, E., IVANESCU, P.L. (HAMMER): "The Transportation Problem with Interchangeable Centers" (in Rumanian). Studii şi Cercetări Matematice, 11,1960,n.2.

/5/. BALAS, E., IVANESCU, P.L. (HAMMER): "A Method for Solving Transportation Problems", (in Rumanian), Comunicările Academiei RPR,11,1961,n.9.

/6/. BALAS, E., IVANESCU, P.L.: "The Parametric Transportation Problem" (in Rumanian), Studii şi Cercetări Matematice, 12, 1961,n.2.

/7/. BALAS, E., IVANESCU, P.L.: "The Transportation Problem with Variable Centers" (in Rumanian), Studii şi Cercetări Matematice, 12,1961, n.2.

/8/. BALAS, E., IVANESCU, P.L.: "Transportation Problems with Variable Data", Revue de Math.pur.appl.,6,1961,n.4.

/9/. BALAS, E., IVANESCU, P.L.: "Transportation of Nonhomogenous Products", (in Rumanian), Studii şi Cercetări Matematice, 13,1962,n.1.

/1o/. BALAS, E., IVANESCU, P.L.: "Stability of Optimal Solutions of Transportation Problems with Changing Costs" (in Rumanian), Comunicǎrile Academiei PBR, 13,1963,n.3.

/11/. BALAS, E., IVANESCU, P.L.: "On the Transportation Problem", Part I, Cahiers du Centre d'Etudes de Recherches Opérationelle, part I,4,.1962, n.2.

/12/. BALAS, E., IVANESCU, P.L.: "On the Transportation Problem", Part II, Cahiers du Centre d'Etudes de Recherche Opérationelle, 4,1962,n.3.

/13/. BALAS, E., IVANESCU, P.L.: "On the Generalized Transportation Problem", Management Science, 11,1964, n.1.

/14/. BEALE, E.M.L.: "A Method of Solving Linear Programming Problems when Some but Not All of the Variables Must Take Integral Values", Statistical Techniques Research Group, Princeton University, Princeton, New Jersey, March 1958.

/15/. BELLMAN, R.: "Dynamic Programming", Princeton University Press, 1957,342.pp.

/16/. BELLMAN, R.E., DREYFUS, S.E.: "Applied Dynamic Programming", Princeton University Press, 1962.

/17/. BEN-ISRAEL, A., CHARNES, A.: "On some Problems of Dyophantine Programming", Cahiers du Centre d'Etudes de Recherche Opérationelle, 4,1962,n.4,pp.215-28o.

/18/. BIRKHOFF, G.: "Lattice Theory", Amer.Math.Soc.Coll.Publ.,25, New York,1948.

/18 bis/. C.BERGE:"Théorie des Graphes et ses Applications", Paris, Dunod,1958.

/19/. CAMION, P.: "Une méthode de résolution par l'algebre de Boole des problèmes combinatoires où interviennent des entiers", Cahiers du Centre d'Etudes de Recherche Opérationelle, 2,196o,n.3.

/2o/. CALDWELL, S.H.: "Switching Circuits and Logical Design", Wiley,1958.

/21/. CHARNES, A., KIRBY, M.: "The Dual Method and the Method of Balas and Ivănescu for the Transportation Problem", Cahiers du Centre d'Etudes de Recherche Opérationelle, vol.6,1964,n.1.pp.5-18-

/22/. DANTZIG, G.B.: "Discrete Variable Extremum Problems", Operations Research, 5, 1957, n.2.

/23/. DANTZIG, G.B.: "On the Significance of Solving Linear Programming Problems with Some Integer Variables", Econometrica, 28.196o,n.1.

/24/. DANTZIG, F.B.: "Linear Programming and Extensions", Princeton University Press, 1963, 621 pp.

/25/. DANTZIG, G.B., HOFFMAN, A.J.: "Dilworth's Theorem on Partially Ordered Sets", Paper n.XI, in "Linear Inequalities and Related Systems", edited by H.W. Kuhn and A.W. Tucker, Ann.Math.Stud.n.38.

/26/. DELEANU, A., IVANESCU, P.L.: "Optimal Assignment of Numbers to Vertices by Pseudo-Boolean Programming", Comm.Coll.on Appl.of Math.in Econ., Jassy (Rumania), 1964.

/27/. DENIS-PAPIN, M., KAUFMANN, A., FAURE, R.: "Cours de Calcul Booléien appliqué", Paris, Albin Michel, 1963.

/28/. DILWORTH, R.P.: "A Decomposition Theorem for Partially Ordered Sets", Annals of Mathematics, 51,195o,pp.161-166.

/29/. EGERVARY, J.: "Matrixok kombinatorikus tulajdonsagairol", Mat.Fiz.Lapok, 1931,p.6.

/3o/. FLAMENT, C.: "Applications of Graph Theory to Group Structure", Prentice-Hall, 1963.

/31/. FLEGG, H.G.: "Boolean Algebra and its Applications", Blackie, 1964.

/32/. FORD, L.R.Jr., FULKERSON, D.R.: "Maximal Flow Through a Network", Canad.J.Math.,8,1956,pp.399-4o4.

/33/. FORD, L.R.Jr., FULKERSON, D.R.: "A Simple Algorithm for Finding Maximal Network Flows and an Application to the Hitchcock Problem", Canad.J.Math.,9,1957,pp.21o-218.

/34/. FORD, L.R.Jr., FULKERSON, D.R.: "Flows in Networks", Princeton University Press, 1962.

/35/. FORTET, R.: "L'algèbre de Boole et ses applications en recherche opérationelle", Cahiers du Centre d'Etudes de Recherche Opérationelle, 1, 1959,n.4.

/36/. FORTET, R.: "Applications de l'algèbre de Boole en Recherche Opérationelle", Revue Française de Recherche Opérationelle, 196o,n.14.

/37/. FORTET, R.: "Résolution Booléenne d'opérations arithmétiques sur les entiers non négatifs et applications aux programmes linéaires en nombres entiers", SMA, Paris, Mars 196o, (Mimeographed).

/38/. FULKERSON D.R.: "Note on Dilworth's Theorem for Partially Ordered Sets", Proc.Amer.Math.Soc.,7,1956,n.4,pp.7o1-7o2.

/39/. FULKERSON, D.R.: "A Network Flow Feasibility Theorem and Combinatorial Applications", Canad.J.Math.,11,1959, pp.440-451.

/40/. FULKERSON, D.R.: "Flows in Networks", in "Recent Advances in Mathematical Programming", editors R.L. Graves and Ph. Wolfe, McGraw-Hill,1963.

/41/. GALE, D.: "A Theorem on Flows in Networks", Pacific J. Math.,7,1957,1073-1082.

/42/. GALE, D.: "The Theory of Linear Economic Models", McGraw-Hill,1960.

/43/. GOMORY, R.: "Essentials for an Algorithm for Integer Solutions to Linear Programming", Bull.Amer.Math.Soc.,64,1958, n.5.

/44/. GOMORY, R.: "Extension of an Algorithm for Integer Solutions to Linear Programming", Amer.Math.Soc.Notices,6, 1959, n.1, Issue, 36, Abstract 553-190.

/45/. HALL, M.Jr.: "Distinct Representatives of Subsets", Bull. Amer.Math.Soc.,54,1948,pp.922-926.

/46/. HALL, M.: "An Algorithm for Distinct Representatives", Amer.Math.Monthly,63,1956,pp.716-717.

/47/. HALL, P.: "On Representatives of Subsets", Jour.London Math. Soc.,10,1935,pp.26-30.

/48/. HALMOS P.R., VAUGHAN, H.E.: "The Marriage Problem", Amer.Jour.Math.,72,1950,pp.214-215.

/49/. HARARY, F., ROSS, I.C.: "A Procedure for Clique Detection Using the Group Matrix", Sociometry, 20, 1957,pp.205-215.

/5o/. HARPER, L.H.: "Optimal Assignment of Numbers to Vertices", Journal of the Soc.Ind.Appl.Math.,12,1964,n.1.pp.131-135.

/51/. HOFFMAN, A.J.: "Some Recent Applications of the Theory of Linear Inequalities to Extremel Combinatorial Analysis", Proc. of Symposia on Applied Math., 1o, 196o,pp.113-128.

/52/. HOFFMAN, A.H., KUHN, H.W.: "Systems of Distinct Representatives and Linear Programming", Amer.Math.Monthly, 63, 1956, pp. 455-46o.

/53/. HOFFMAN, A.J., KUHN, H.W.: "On Systems of Distinct Re-Representatives", Annals of Math. Studies, n.38,1956, pp.199-2o6.

/54/. IVANESCU, P.L. (HAMMER): "Shefferian Algebras", Bull. Math. Soc.Sci.Math.Phys.RPR, 3(51),1959,n.4.

/55/. IVANESCU, P.L.: "Applications of Galois Fields to Discrete Mathematical Programming", Comm. at the Coll.Approx.Funct. Appl.Num.Anal.,Cluj, November,15-19,1963.

/56/. IVANESCU, P.L.: "Programmation polynomiale en nombres entiers" C.R.Acad.Sci.Paris,258,1964,n.2.

/57/. IVANESCU, P.L.: "Some Network-Flow Problem Solved with Pseudo-Boolean Programming",Operations Research. 13. 1965. n.3.

/58/. IVANESCU, P.L.: " On the Minimal Decomposition of Finite Partially Orderes Sets in Chains", Rev.Roum.Math.Pur.Appl.9, 1964,n.1o

/59/. IVANESCU, P.L.: "Systems of Pseudo-Boolean Equations and Inequalities", Bull.Acad.Polon.Sci.,Cl.III,12,1964,n.11

/6o/. IVANESCU, P.L.: "The Method of Succesive Eliminations for Pseudo-Boolean Equations", Bull.Acad.Polon.Sci.Cl.III,12, 1964,n.11

/61/. IVANESCU, P.L.: "The Reduced of a Pseudo-Boolean Function", Bull.Acad.Polon.Sci.Cl.III, 12. 1965. n.11

/62/. IVANESCU, P.L.: "The Pseudo-Boolean Method for Integer Polynomial Programming", Bull.Acad.Polon.Sci.Cl.III 12. 1965.n.11.

/63/. IVANESCU, P.L.: "Integer Polynomial Programming", Comm. at the International Symposium on Mathematical Programming. London School of Economics, London, July 6-1o,1964. Mimeographed.

/64/. IVANESCU, P.L.: "Sur la résolution par l'algèbre de Boole des problèmes de programmation polynomiale en nombres entiers", C.R.Acad.Sci.Paris, (to appear).

/65/. IVANESCU, P.L.: "Dynamic Programming with Discrete Bivalent Variables", Comn.Coll.. Appl. Math. in Economies. Smolenice. Czechoslovakia. June. 1965

/66/. IVANESCU, P.L.: "Accelerating the Minimization of Pseudo-Boolean Functions", Manuscript.

/67/. IVANESCU, P.L.: "Sur le couplage des graphes simples", Comm. Coll. Appl. of Convex Functions. Cluj. Rumania. July. 1965

/68/. IVANESCU, P.L.: "Systems of Distinct Representatives", Manuscript.

/69/. IVANESCU, P.L. and ROSENBERG, I.: "Applications of Pseudo-Boolean Programming to the Theory of Graphs", Zeitschrift für Wahrscheinlichkeitstheorie, 3 , 1964,n.2.

/7o/. IVANESCU, P.L., ROSENBERG, I. and RUDEANU, S.: "On the Minimization of Pseudo-Boolean Functions", (in Rumanian), Studii şi Cercetări Matematice,14,1963,n.3.

/71/. IVANESCU, P.L., ROSENBERG, I. and RUDEANU, S.: "Pseudo-Boolean Programming", Proc.Coll.Appl.Math.in Econ., Budapest, June 18-22,1963 (to appear).

/72/. IVANESCU, P.L., ROSENBERG, I. and RUDEANU, S.: "Application of Discrete Linear Programming to Minimization of Boolean Functions" (in Russian), Rev.Math.Pur.Appl.,8,1963,n.3.

/73/. IVANESCU, P.L. (HAMMER) and RUDEANU, S.: "On the Egervary Method for Solving Transportation Problems, I", (in Rumanian), Comunicările Academiei RPR, 11, 1961,n.7.

/74/. IVANESCU, P.L. and RUDEANU, S.: " On the Egervary Method for Solving Transportation Problems,II", (in Rumanian), Studii şi Cercetări Matematice, 14, 1963, n.1.

/75/. IVANESCU, P.L. and RUDEANU, S.: "An Application of Boolean Algebra to the Transportation Problem", Comm.at Coll. Math. Logics and Appl., Tihany (Hungary), September, 1962.

/76/. IVANESCU, P.L. and RUDEANU, S.: "Minimization of Switching Circuits in Case of Hazard", Circullar Letter N.14 IFAC Symposium on Race and Hazard Phenomena in Switching Circuits, Bucharest 1965

/77/. JOHNSON, W.E.: "Sur la théorie des équations logiques", Bibl. Congr.Internat.Phil.,III,19o1.

/78/. KOLUMBAN, I.: "On the Egervary Method for Solving the Transportation Problem", Comm. at the Coll. on Numer.Anal.,Cluj, Rumania, December 1962.

/79/. KÖNIG, D.: "Theorie der Endlichen und Unendlichen Graphen", Leipzig, 1936.

/8o/. KÖNIG, D., VALKO, S.: "Über mehrdeutige Abbildungen von Mengen", Math. Annalen, 95, 1926, p.135.

/81/. KUHN, H.W.: "The Hungarian Method for the Assignment Problem", Naval Research Logistics Quarterly, 2,1955,n.2.pp.83-97.

/82/. KUHN, H.W.: "Variants of the Hungarian Method for Assignment Problems", Naval Research Logistics Quarterly,3,1956, pp.253-258.

/83/. KÜNZI, H.P. and OETTLI, W.: "Integer Quadratic Programming", in "Recent Advances in Mathematical Programming" (ed.R.L. Graves and Ph. Wolfe), McGraw-Hill Book Company, 1963, pp.303-308.

/84/. KREKO, B.: "Linearis Programmozas", Budapest, 1962.

/85/. LAND, A.H. and DOIG, A.G.: "An Automatic Method of Solving Discrete Programming Problems", London School of Economics, 1957.

/86/. LÖWENHEIM, L.: "Über die Auflösung von Gleichungen im Logischen Gebietkalkül", Math. Ann., 68, 1910.

/87/. LUCE, R.D.: "Connectivity and Generalized Cliques in Sociometric Group Structure", Psychomet, 15, 1950,pp. 169-190.

/88/. McCLUSKEY, E.J.: "Minimization of Boolean Functions", Bell. System Techn.J.,35,1956,pp.1417-1444.

/89/. MAGHOUT, K.: "Applications de l'Algèbre de Boole à la théorie des graphes et aux programmes lineaires et quadratiques", Cahiers du Centre d'Etude de Recherche Opérationelle, 5, 1963,n.1-2.

/90/. MAISTROVA, T.L.: "Linear Programming and the Problem of Symplification of Normal Forms of Boolean Functions" (in Russian) Probl.Peradaci Informatii, n.12,pp.5-15, Moscow, 1962.

/91/. MANN, H.B., RYSER, H.J.: "Systems of Distinct Representatives", Am.Math.Monthly, 6o, 1953,p.397.

/92/. MENDELSOHN, N.S., DULMAGE, A.L.: "Some Generalizations of the Problem of Distinct Representatives", Canad.J.Math.,1o, 1958,pp.23o-241.

/93/. MOISIL, Gr.C.: "Scheme cu Comandă Directă cu Contacte şi Relee", Editura Academiei RPR, Bucureşti, 196o.

/94/. MOISIL, Gr.C.: "Funcţionarea în mai mulţi Timpi a Schemelor cu Relee Ideale", Editura Academiei RPR, Bucureşti, 1961.

/95/. MOISIL, Gr.C.: "Circuite cu Tranzistori", vol.I,II, Editura Academiei RPR, Bucureşti 1963.

/96/. MOISIL, Gr.C.: "Algebraic Theory of Switching Circuits" (in Rumanian: Bucureşti, 1959; in Russian: Moscow, 1963; English translation in press).

/97/. MOISIL, Gr.C.: "An Algebraic Theory of the Actual Operation of Relay Switching Circuits", IFAC Symp. on Hazard and Race Phenomena in Switching Circuits, Circular Letter No. 5, Bucharest 1964.

/98/. NEMETI, L., RADO, F.: "Ein Wartezeitproblem in der Programmierung der Production", Mathematica (Cluj), (in press).

/99/. ORE, O.: "Graphs and Matching Theorems", Duke Math.J.,22,1955.

/1oo/. ORE, O.: "Theory of Graphs", Amer.Math.Soc.Colloq.Publs.,38, 1962.

/1o1/.QUINE, W.V.: "The Problem of Simplifying Truth Functions", Trans.Amer.Math.Monthly,59,1952,pp.521-531.

/1o3/.RADO, F.: "Linear Programming with Logical Conditions", (in Rumanian), Comunicările Academiei RPR, 13, 1963,n.12.pp. 1o39-1o42.

/1o4/. RADO, R.: "Factorisation of Even Graphs", Quarterly Jour. Math.,2o,1949,pp.95-1o4.

/1o5/.ROTH, J.P.: "Algebraic Topological Methods for the Synthesis of Switching Systems", Trans.Amer.Math.Soc.,88,1958, pp.3o1-326.

/1o6/.ROTH, J.P.: "Algebraic Topological Methods in Synthesis", Proc.Internal,Symp.on the Theory of Switching, part.II, Harvard University, Cambridge, April 1957, pp.57-73.

/1o7/.ROTH, J.P., WAGNER, E.G.: "Algebraic Topológical Methods for the Synthesis of Switching Systems, Part III: Minimization of Non singular Boolean Trees", IBM Journal of Research and Development, 3,1959,n.4.,pp.326-344.

/1o8/.RUDEANU, S.: "Boolean Equations and their Applications to the Study of Bridge-Circuits,I", Bull.Math.,3(51),1959, n.4.

/1o9/.RUDEANU, S.: "Boolean Equations and their Applications to the Study of Bridge-Circuits, II", (in Rumanian), Comunicările Academiei RPR,11,1961,n.6.

/11o/.RUDEANU, S.: "On Solving Boolean Equations with the Method of Löwenheim", (in Rumanian), Studii şi Cercetări Matematice, 13,1962,n.2.

/111/.RYSER, H.J.: "Combinatorial Mathematics", The Carus Math. Monographs, n.14,1963.

/112/.SIMMONNARD, G.: "Programmation linéaire", Paris, Dunod, 1962.

/113/.SZASZ, G.: "Introduction to Lattice Theory", Academy, Budapest, 1963.

/114/.URBANO, R.H., MUELLER, R.K.: "A Topological Method for the

Determination of the Minimal Forms of a Boolean Function", 5, 1956, pp.126-132.

/115/.YABLONSKI, S.V.: "Functional Constructions in k-valued Logics", (Russian), Trudy Mat.Inst. im V.A.Steklova, n.51,1958, pp.5-142.